Table of Contents

AF894685

Chemistry Notables

Schedule and Due Dates

Time / Period	Monday	Tuesday	Wednesday	Thursday	Friday

Assignment / Project / Exam	Due Date

Assignment / Project / Exam	Due Date

Chemistry Notables

1 Periodic Table

A table of the chemical elements arranged by atomic number.

Key: Actinides · Alkali metals · Alkaline earth metals · Lanthanides · Metalloids · Noble gases · Post-transition metals · Reactive nonmetals · Transition metals · Unknown properties

1	2	3	4	5	6	7	8	9	10	11	12	13	14	15	16	17	18
1 H																	2 He
3 Li	4 Be											5 B	6 C	7 N	8 O	9 F	10 Ne
11 Na	12 Mg											13 Al	14 Si	15 P	16 S	17 Cl	18 Ar
19 K	20 Ca	21 Sc	22 Ti	23 V	24 Cr	25 Mn	26 Fe	27 Co	28 Ni	29 Cu	30 Zn	31 Ga	32 Ge	33 As	34 Se	35 Br	36 Kr
37 Rb	38 Sr	39 Y	40 Zr	41 Nb	42 Mo	43 Tc	44 Ru	45 Rh	46 Pd	47 Ag	48 Cd	49 In	50 Sn	51 Sb	52 Te	53 I	54 Xe
55 Cs	56 Ba	57 La	72 Hf	73 Ta	74 W	75 Re	76 Os	77 Ir	78 Pt	79 Au	80 Hg	81 Tl	82 Pb	83 Bi	84 Po	85 At	86 Rn
87 Fr	88 Ra	89 Ac	104 Rf	105 Db	106 Sg	107 Bh	108 Hs	109 Mt	110 Ds	111 Rg	112 Cn	113 Nh	114 Fl	115 Mc	116 Lv	117 Ts	118 Og
			58 Ce	59 Pr	60 Nd	61 Pm	62 Sm	63 Eu	64 Gd	65 Tb	66 Dy	67 Ho	68 Er	69 Tm	70 Yb	71 Lu	
			90 Th	91 Pa	92 U	93 Np	94 Pu	95 Am	96 Cm	97 Bk	98 Cf	99 Es	100 Fm	101 Mn	102 No	103 Lr	

2 Element List

1. Hydrogen	2. Helium	3. Lithium	4. Beryllium	5. Boron
6. Carbon	7. Nitrogen	8. Oxygen	9. Fluorine	10. Neon
11. Sodium	12. Magnesium	13. Aluminium	14. Silicon	15. Phosphorus
16. Sulfur	17. Chlorine	18. Argon	19. Potassium	20. Calcium
21. Scandium	22. Titanium	23. Vanadium	24. Chromium	25. Manganese
26. Iron	27. Cobalt	28. Nickel	29. Copper	30. Zinc
31. Gallium	32. Germanium	33. Arsenic	34. Selenium	35. Bromine
36. Krypton	37. Rubidium	38. Strontium	39. Yttrium	40. Zirconium
41. Niobium	42. Molybdenum	43. Technetium	44. Ruthenium	45. Rhodium
46. Palladium	47. Silver	48. Cadmium	49. Indium	50. Tin
51. Antimony	52. Tellurium	53. Iodine	54. Xenon	55. Caesium
56. Barium	57. Lanthanum	58. Cerium	59. Praseodymium	60. Neodymium
61. Promethium	62. Samarium	63. Europium	64. Gadolinium	65. Terbium
66. Dysprosium	67. Holmium	68. Erbium	69. Thulium	70. Ytterbium
71. Lutetium	72. Hafnium	73. Tantalum	74. Tungsten	75. Rhenium
76. Osmium	77. Iridium	78. Platinum	79. Gold	80. Mercury
81. Thallium	82. Lead	83. Bismuth	84. Polonium	85. Astatine
86. Radon	87. Francium	88. Radium	89. Actinium	90. Thorium
91. Protactinium	92. Uranium	93. Neptunium	94. Plutonium	95. Americium
96. Curium	97. Berkelium	98. Californium	99. Einsteinium	100. Fermium
101. Mendelevium	102. Nobelium	103. Lawrencium	104. Rutherfordium	105. Dubnium
106. Seaborgium	107. Bohrium	108. Hassium	109. Meitnerium	110. Darmstadtium
111. Roentgenium	112. Copernicium	113. Nihonium	114. Flerovium	115. Moscovium
116. Livermorium	117. Tennessine	118. Oganesson		

Chemistry Notables Level 1

3 **Intro to Chemistry** **Science**- a branch of knowledge or study dealing with a body of facts or truths systematically arranged and showing the operation of general laws.
Chemistry- the science that deals with the composition and properties of substances and various elementary forms of matter.
Chemical- a substance produced by or used in a chemical process.

4 **Atom** smallest unit of matter. (Level 2) the smallest component of an element having the chemical properties of the element, consisting of a nucleus containing combinations of neutrons and protons and one or more electrons bound to the nucleus by electrical attraction; the number of protons determines the identity of the element.

5 **Atomic Number** the number of positive charges or protons in the nucleus of an atom of a given element.

6 **Bond** An attraction between atoms that forms chemical substances.

7 **Compound** a pure substance composed of two or more elements whose composition is constant.

8 **Electron** negatively charged particle that orbits the nucleus of an atom.

9 **Element** one of a class of substances that cannot be separated into simpler substances by chemical means.

10 **Ion** an electrically charged atom or group of atoms formed by the loss or gain of one or more electrons.

11 **Isotope** any of two or more forms of a chemical element, having the same number of protons in the nucleus, or the same atomic number, but having different numbers of neutrons in the nucleus, or different atomic weights.

12 **Mass Number** the integer nearest in value to the atomic weight of an atom and equal to the number of nucleons in the nucleus of the atom.

13 **Matter** the substance of which a physical object is composed.

14 **Metal** a substance that is a good conductor of heat and electricity. They are generally malleable, ductile, and shiny.

15 **Mixture** A combination of two or more substances that are not chemically combined.

16 **Molecule** the smallest physical unit of an element or compound, consisting of one or more like atoms in an element and two or more different atoms in a compound.

17 **Neutron** an elementary particle having no charge, mass slightly greater than that of a proton.

18 **Solution** a homogeneous, molecular mixture of two or more substances. (Level 2) the process by which a gas, liquid, or solid is dispersed homogeneously in a gas, liquid, or solid without chemical change.

19 **Solvent** a substance that dissolves another to form a solution.

Chemistry Notables Level 2

20 **Acid** a compound having a pH value of less than 7, usually having a sour taste and capable of neutralizing alkalis and reddening blue litmus paper.

21 **Alkali metals** any of the group of univalent metals including potassium, sodium, lithium, rubidium, cesium, and francium, whose hydroxides are alkalis.

22 **Alkaline earth metals** the group of bivalent metals including barium, radium, strontium, calcium, and, usually, magnesium, the hydroxides of which are alkalis but less soluble than those of the alkali metals.

23 **Allotropes** one of two or more existing forms of an element: Graphite and diamond are allotropes of carbon.

24 **Alloy** a substance composed of two or more metals, or of a metal or metals with a nonmetal, intimately mixed, as by fusion or electrodeposition.

25 **Base** is the opposite of acid and has a pH of 7 to 14. Compounds that form hydroxyl ions (OH) when dissolved in water, and whose aqueous solutions have a bitter taste and react with acids to form salts. Bases turn red litmus paper blue.

26 **Chemical reaction** A process where a set of substances undergo a chemical change to form a different substance.

27 **Covalent bond** occurs between atoms that share a pair of electrons in order to gain full outer shells.

28 **Density** Density is the mass per unit volume of a substance.

29 **Distillation** The process of separating mixtures by boiling a liquid and then recovering the cooled gas.

30 **Electrolysis** A process used to drive a chemical reaction by using an electrical current. It is often used to separate elements.

31 **Noble gases** A family of elements that makes up column 18 of the periodic table. They are very stable due to their full outer electron shells.

32 **Oxidation** The name for the process when oxygen combines with another element to form a compound.

33 **PH scale** measures the activity of the hydrogen ion in substances. If a substance has a pH less than 7 it is said to be acidic. If its PH is greater than 7 then it said to be basic or alkaline.

34 **Radioactivity** or radioactive decay is a process in which an unstable nucleus loses energy by the emission of radiation in the form of a particle.

35 **Reaction** The process that occurs when two or more chemical substances are combined to form another chemical substance.

36 **Salts** are chemical compounds that are formed by the reaction of an acid and a base.

Made in the USA
Monee, IL
05 November 2026

67385764R00045